NEXT-GENERATION MILITARY TECHNOLOGY

CYBER WARFARE
IN THE MILITARY

by Cynthia Kennedy Henzel

BrightPoint Press

San Diego, CA

BrightPoint Press

an imprint of ReferencePoint Press, Inc.
Printed in the United States

For more information, contact:
BrightPoint Press
PO Box 27779
San Diego, CA 92198
www.BrightPointPress.com

LIBRARY OF CONGRESS CATALOGING-IN-PUBLICATION DATA

Name: Henzel, Cynthia Kennedy, author.
Title: Cyber warfare in the military / by Cynthia Kennedy Henzel.
Description: San Diego, CA: ReferencePoint Press, 2027 | Series: Next-generation military technology | Includes bibliographical references and index. | Audience: Grades 7–9
Identifiers: ISBN 9781678214227 (hardcover) | ISBN 9781678214234 (eBook)
The complete Library of Congress record is available at www.loc.gov.

CONTENTS

AT A GLANCE

- Cyber warfare is the use of computers and the internet to disrupt other countries.
- A computer program called Stuxnet was the first major cyber weapon.
- North Korea, Russia, Iran, and China all engage in cyberattacks.
- Countries use cyber warfare to steal money for their military.
- Cyber warfare is used to spy on government officials and steal data.
- Countries use cyber warfare to disrupt or destroy electricity grids and water treatment plants.

- One form of cyber warfare involves posting misinformation on social media to disrupt elections.
- Keeping computer systems updated with the latest software helps defend against cyberattacks.
- Teaching users how to protect themselves from common methods that people use to break into computer systems helps prevent cyberattacks.
- Powerful US cyber warfare capabilities may deter other countries from attacking the United States.

OPERATION GLOWING SYMPHONY

In 2014, a group called the Islamic State of Iraq and Syria (ISIS) began terror attacks around the world. Its goal was to spread extreme Muslim views. ISIS used online videos to share its ideas. It also recruited new members online.

The United States government tracked ISIS online. It was slow, hard work. Many of the group's sites were **encrypted**. Then a young marine noticed something.

Cyber warfare operators monitor cyberattacks from Warfield Air National Guard Base in Middle River, Maryland. The operations floor is known as the Hunter's Den.

FY18 Wing Goals

ISIS mostly used the same ten internet accounts. These accounts were for banking and publishing online.

The US Cyber Command worked with the National Security Agency (NSA). Together, they came up with a plan. It was called Operation Glowing Symphony. The goal was to take down ISIS online. They found the passwords of ISIS members. This helped them get into accounts. They noticed which apps ISIS account owners used. This helped them learn what the ISIS members were doing.

The team sent out a **phishing** email. ISIS members clicked on the link. This let the US team sneak into the ISIS computers. They installed malware. This software harms computer systems. They also

In Aleppo, Syria, local military convoys fight ISIS strongholds on the ground.

searched for information. They found attack plans. They also found the names of ISIS members.

ISIS used large **servers**. The servers also stored information from other organizations. These included hospitals. Destroying the servers would destroy all data. The cyber warriors had to be careful. They practiced for months to target only ISIS.

On attack day, the team hit all ten ISIS accounts. They began destroying files. Suddenly, the attack stopped. A security question appeared on the screen. "What is the name of your pet?" An analyst stood up. "Sir, 1-2-5-7," the analyst said. "He [uses] it for everything."[1] They typed in the numbers. They were in.

Security questions are a common method used to add an extra layer of protection beyond passwords.

The attack worked. The ISIS media network crumbled. Its websites disappeared. ISIS was seriously damaged.

CYBER WARFARE

To stop ISIS, the United States used cyber warfare. This means using computers to disrupt the actions of a country or organization. Some cyberattacks steal information. Others steal money from banks or businesses. Some attacks take down electrical grids.

Cyber warfare is also used to spread false information. Cyberattacks can make citizens mistrust their governments or other citizens. The United States needs more cyber experts to protect the nation from cyber warfare.

A NEW WAY TO FIGHT

Internet technology has brought a major shift in warfare. It allows small nations or terrorist groups to launch cyberattacks. The first major cyberattack happened in 1988. At that time, only university researchers used the internet. Robert Morris was a student at Cornell University. As an experiment, he created a worm. A worm is a type of malware.

Cyber warfare between nations is expected to increase significantly in the years ahead.

TARGET: 55°45'6"N 37°37'4"E
TARGET: 48°51'24"N 2°21'08"E

The Morris Worm did not destroy files. But it slowed down computers. Victims included the National Aeronautics and Space Administration (NASA). It affected major universities, too.

The Morris Worm cost the victims hundreds of thousands of dollars. Universities had to shut down computers to remove the worm. Morris had not meant to cause harm. He provided instructions for removing the worm. The Morris Worm showed the danger of hacking. Hacking is illegally breaking into a computer.

EARLY CYBERATTACKS

The World Wide Web was created in the early 1990s. This service lets people access websites using the internet. More and more

Robert Morris launched the Morris Worm from the Massachusetts Institute of Technology to avoid having it traced back to him.

people began using the internet. By the mid-1990s, millions of people were online.

Early cyberattacks often targeted email systems. Most were phishing attacks. Users were tricked into giving out passwords or other information. Sometimes, they were asked to click on a link. The link could give worms and viruses access to the user's computer. In 1999, 15-year-old Jonathan James hacked into computers at the US Department of Defense. He found usernames and passwords. He used these to hack into NASA software. The software controlled the International Space Station (ISS). It took 21 days for NASA to repair the systems. Jonathan explained he was just playing around. He had not meant to cause trouble.

In 2000, 15-year-old Michael Calse launched cyberattacks against large companies. These included Amazon and eBay. He used distributed denial of service (DDoS) attacks. DDoS attacks flood networks with requests. Too many requests stress the network. This forces it to crash.

MILITARY HACKING

Governments soon realized the power of cyber warfare. John Arquilla and David

Jonathan James hacked into NASA's computer system in late 1999, just months after *Discovery* became the first space shuttle to dock with the ISS.

Ronfeldt were researchers. In 1993, they predicted that countries would fight future wars through computer networks. They wrote, "Whoever masters the network form will gain major advantages."[2]

The first major use of cyber warfare was the Stuxnet worm. The United States and Israel began this cyberattack in 2007. The malware damaged Iranian computer systems. These were part of Iran's nuclear program. Stuxnet destroyed **centrifuges** by making them spin out of control.

After Stuxnet, other countries began similar cyberattacks. Some used these methods to attack the United States. Kim Zetter wrote a book about Stuxnet. She says, "When you launch a cyber weapon, . . . you send the [information] that created it and the ability to launch the weapon back against you."[3]

HYBRID WARFARE

Georgia is a country that borders Russia. In 2008, Georgia started working more closely with Europe. Russia did not like this. In August, Russia attacked the networks of the Georgian government. It used DDoS attacks. It hacked into Georgian media websites. It used these sites to give false information to confuse the public.

Gas centrifuges are used to produce enriched uranium. This material is needed to generate nuclear power and build nuclear weapons.

Russia used malware to take information from Georgian systems.

Russia then launched missiles against Georgia on August 1. Georgia tried to **negotiate**. But talks failed. Georgia began fighting back on August 7. Russia's military quickly overpowered Georgia's. The countries agreed to a ceasefire on August 13. The ceasefire made two

Russia Wants More

Russia's hybrid war with Georgia succeeded. This encouraged more attacks. In 2014, Russia launched a cyber war against Ukraine. Russia then occupied a Ukrainian territory called Crimea. In 2015, Russia began new cyberattacks. It targeted energy and banking systems in Ukraine. These attacks increased just before Russia invaded Ukraine in 2022.

northern regions of Georgia independent. Russia soon occupied the territories. This was a **hybrid** approach to cyber warfare. Russia used cyberattacks to confuse and weaken Georgia. Then it used physical attacks to gain Georgian territory.

In addition to hacking into Georgian websites, Russia sent troops to the country of Georgia.

SPYING AND THEFT

Cyber warfare is a threat to the United States. Four main countries use cyber warfare. These are North Korea, Russia, China, and Iran. Some of these countries have government hacking groups. Others pay outside hacking groups for their work.

One famous hacking group is the Lazarus Group. It works for North Korea. In 2014, the Lazarus Group attacked Sony Pictures. Sony had made a movie called

Kim Jong-Un is the third supreme leader from the Kim family to lead North Korea. He took power in April 2012.

The Interview. This comedy made fun of Kim Jong-Un. He is the leader of North Korea. An official in North Korea said the movie was "an act of terrorism."[4] The cyberattack was revenge for making fun of their leader.

The cyberattack began before the movie was released. The Lazarus Group sent out phishing emails. Sony employees clicked

Hacktivism

Hacktivists break into the computer systems of organizations as a way to protest them. They hack companies that cause damage to the environment or public health. They target politicians for supporting certain policies. Anonymous is a famous hacktivist organization. It makes hidden information public. Anonymous does this to try to bring about political change.

on the attachments. The phishers got in. They soon spread malware through the Sony network.

The Lazarus Group stole data. It also stole movies that Sony had not released. The Lazarus Group posted embarrassing Sony emails on the internet. It put the company's movies on the internet. People could download them for free. The Lazarus Group threatened theaters that showed *The Interview*. Many theaters feared terrorist attacks. They did not show the movie. Sony decided to show the movie in only a few places.

North Korea is a poor country. It has used cyber warfare to steal money for its military. In 2016, the Lazarus Group hacked the Bangladesh Bank. Again, the group used phishing. One employee opened an email and was tricked into downloading malware. The Lazarus Group was in. It stole $81 million.

North Korea also uses ransomware. This is another type of malware used for cyber warfare. Hackers break into a system

Encryption scrambles readable text into code. This allows sensitive data to be transferred safely.

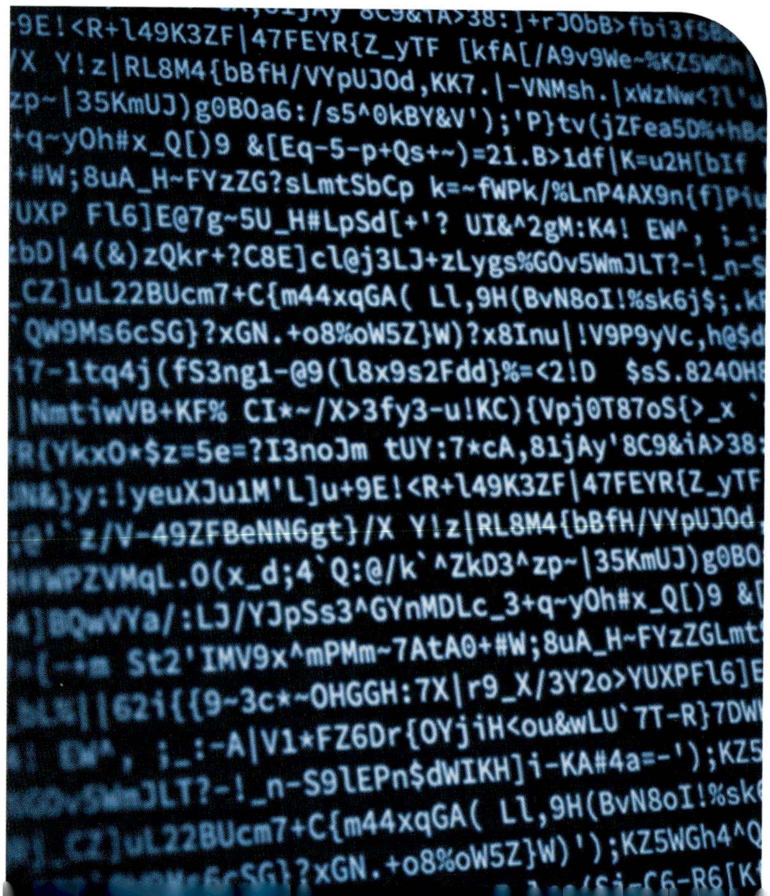

and encrypt the data. The legal owners of the data cannot read it. The hackers then demand money. If they get the money, they provide a key to decode the data.

In 2017, the Lazarus Group began the WannaCry ransomware attack. The attack affected more than 200,000 computers. This impacted 150 countries. Many of the computers were in hospitals. Others were in transportation systems. Many owners paid the ransom. North Korea got billions of dollars for its military.

RUSSIAN HACKING

Cozy Bear is a hacking group working with Russia. It is famous for spying. In 2014, Cozy Bear hacked into many US government email systems.

These included the State Department and the White House. Cozy Bear used phishing to get into the systems. It also used malware. It then installed backdoors. Backdoors are secret ways to get into a system. They work around the system's security. Cozy Bear gave Russia private messages between top-level US officials.

In 2019, Russian spies used a new method to steal data. SolarWinds is a company. It makes a software called Orion. The US government and many companies use it. When users saw an update from SolarWinds, they clicked on it. Updates are used to fix software bugs. Bugs are small errors in software. Updates also provide new features. But this time, hackers used the update to get into the

SolarWinds was founded in 1999 in Austin, Texas. The company helps businesses manage their computer networks and systems.

system. Christopher Krebs was in charge of protecting government systems. He says, "[The Russian spies] move like ghosts. They are very hard to track."[5]

SolarWinds users did not notice anything wrong. So, the Russians stole data for almost a year. This included data from the US Cybersecurity & Infrastructure Security Agency (CISA).

The SolarWinds attack showed that cyberattacks against private companies can affect the US government. The US government works with thousands of private companies. They provide accounting services. They build weapons. They support the military. US defense depends on protecting private US companies.

In 2021, a group called DarkSide was working for Russia. DarkSide noticed something. Colonial Pipeline is a fuel company. It supplies fuel to the US East Coast. DarkSide discovered that Colonial Pipeline had not updated its software. This allowed DarkSide to hack the system. It installed ransomware. The DarkSide attack caused fuel shortages on the East Coast for more than a week. Colonial Pipeline paid a $5 million ransom. But it still had to shut down the pipeline to prevent further damage. This incident showed that cyberattacks could damage US electrical grids and other systems.

Following cyber security best practices is key to preventing cyberattacks. These include using strong passwords and updating software regularly.

CHAPTER THREE

CYBER WARFARE INCREASES

In 2010, Google announced a major attack from China. The company called it Operation Aurora. Hackers got into Google accounts. Some accounts showed legal **wiretaps** ordered by the US government. China used this data to find out which Chinese spies had been discovered by the United States. China could warn its spies to leave the country.

Google's leadership and engineering teams are based at the company's headquarters in Mountain View, California.

Operation Aurora also stole company secrets. This included plans for new weapons and space technology. Operation Aurora attacked thirty-four US companies. China used the secrets to improve its own technology.

China has a well-funded cyber warfare program. It uses advanced technology.

China has sometimes used hackers to learn about technology used in US military aircraft. From 2008 until 2014, hackers working for China's army stole files about the C-17, the F-35, and the F-22 Raptor (pictured).

Its hackers are secretive. Chinese malware called BRICKSTORM has been found in US government agency computer systems. It allows long-term access to infected computers.

China is also planning future cyberattacks. In 2025, the Salt Typhoon cyberattack targeted communications companies. The Volt Typhoon attack targeted infrastructure systems. These included electric grids and water treatment plants. These attacks did not shut down the systems. They were used to place malware in computers. The malware could be used if needed in the future. Christopher Wray was the director of the FBI. He said, “China’s hacking program is larger than that of every other major nation, combined.”[6]

ELECTIONS AND SOCIAL MEDIA

Cyber warfare has become part of elections. It is used to influence voters. It creates mistrust in election results. Enemies also use cyber warfare to divide Americans.

In 2016, Russia launched many cyberattacks. It wanted to target elections in the United States. Russia wanted Donald Trump to win the 2016 presidential election. He was running against democrat Hillary Clinton. Cozy Bear worked with Fancy Bear, another Russian hacking group. They broke into the Democratic National Committee (DNC) computer system. The hackers took private information. They posted it online. They wanted to harm Clinton's campaign.

Hackers also used disinformation. This is false information that is used to

In 2016, Hillary Clinton won the popular vote by 48 percent to Trump's 46 percent. But Trump won the electoral college and the election.

mislead people. Computer **bots** and Russian officials pretended to be Americans online. They organized protests using social media. Many Americans believed the protests were led by real people. Russia tried to hack into voting systems in twenty-one states. They were not able to change votes. But they did get voter registration information. Trump won the election. It is not known if the Russian cyberattacks influenced the result.

THE 2020 ELECTION

General Paul Nakasone took charge of the NSA in 2018. He reviewed the 2016 Russian attacks. "It provided us with a very, very good road map of what they might do in the future," said Nakasone.[7]

The 2020 election was between President Trump and Joe Biden. This time, the United States was better prepared for cyberattacks. Most states had paper ballots. These could be counted if cyberattacks damaged voting machines. States updated voting security systems. In 2020, many people could not go to the polls to vote. This was due to the

During the 2020 presidential election, 69 percent of voters cast their ballot by mail or voted at the polls before election day.

During the COVID-19 pandemic, 42 percent of voters used mail-in voting.

COVID-19 pandemic. So states used more mail-in voting. They made voting periods longer. Misinformation led some people to think this was unfair.

Russian groups carried out more than 200 cyberattacks against US political groups. China sponsored cyberattacks against the Biden campaign and other international leaders.

The United States used its own cyberattacks to damage known hackers. This helped it stop many attacks quickly. Still, misinformation destroyed the trust of many citizens. Biden won the election. But some people refused to believe that he won.

Cyberattacks Can Fail

In August 2024, Iranian hackers broke into Trump's campaign computer systems. They stole information. They then offered the information to the Biden campaign. The Democrats ignored the emails. The hackers also sent the stolen information to media such as the *New York Times* and *Washington Post*. The media also ignored the emails. The cyberattack failed.

2024 ELECTION

After 2020, artificial intelligence (AI) advanced quickly. It became difficult to tell if an image or video was real or made by AI. AI was soon used for cyber warfare in elections. AI chatbots made fake posts. But the posts looked as if they were made by real people. Bots used **deepfake** videos. These make it look as if a trusted person is telling a fake story. That person might be a reporter. Bots created deepfakes of famous people supposedly supporting a policy.

In one deepfake, a Republican supporter used Biden's voice for fake calls. The fake voice told people that he was so far ahead that they did not need to vote. Jessica Rosenworcel headed the Federal Communications Commission.

In the 2024 presidential election, Trump defeated Democratic candidate Kamala Harris, who ran after Biden withdrew from the race.

This agency ensures that communication by electronic devices follows the law. Rosenworcel explains that AI makes it easy to create deepfakes.

Trump won the 2024 presidential election. Cyber warfare continued. President Trump took office in January 2025. In March 2025, he began firing many federal workers. China used this event to launch a cyber warfare campaign. It sent out fake ads for jobs. Some of the fired employees answered the ads. China tried to get secret information from them.

China also launched the first AI-controlled cyberattack in 2025. It used AI instead of a human hacking team. AI is much faster than a human team. It can also work longer. And AI is able to scan huge data systems.

COUNTRIES THAT LAUNCH POLITICAL CYBERATTACKS, 2000–2023

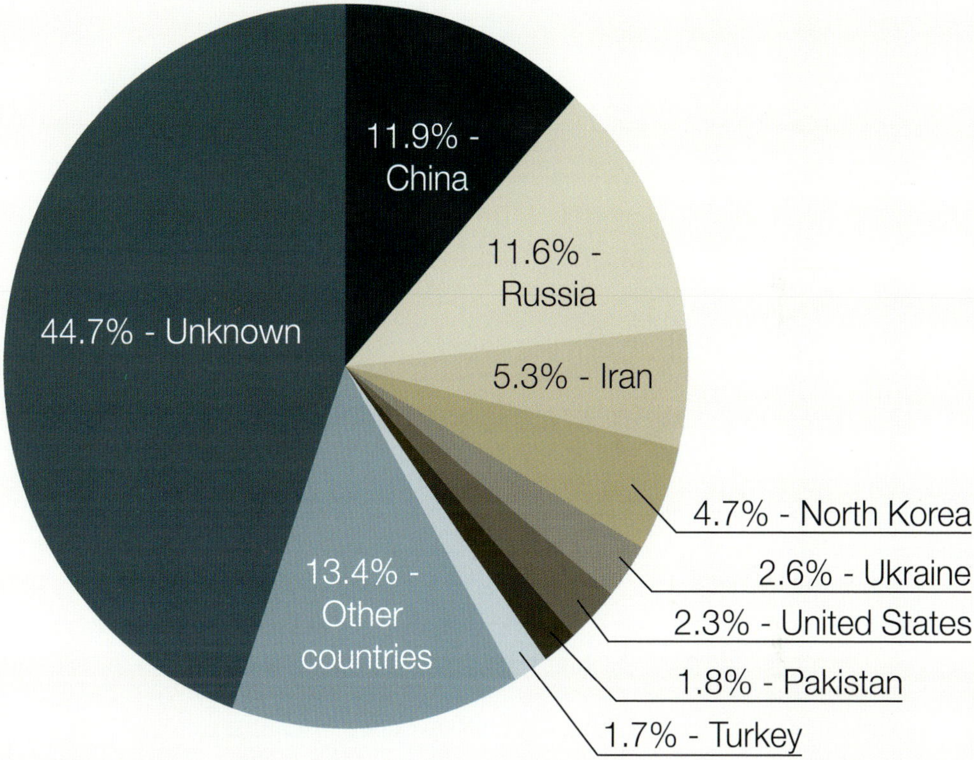

Source: Anna Fleck, "Who's Behind Cyber Attacks?" Statista, *February 23, 2024. www.statista.com.*

The European Repository of Cyber Incidents defined political cyberattacks as those that had political objectives or were directed toward political targets or critical infrastructure.

CHAPTER FOUR

WINNING THE CYBER WAR

In 2004, US military leaders declared that cyberspace was a place for warfare. In the past, wars took place in the air, on land, at sea, and in space. Now war is also waged in cyberspace. The US Navy and Army created new cyber warfare units. In May 2010, the United States created the US Cyber Command (Cybercom). Cybercom brings together cyber warfare experts from all military branches.

Cyber command units for different branches of the US military focus on many missions, including spying, electronic warfare, and protection against cyberattacks.

In 2012, Cybercom and military leaders created the Cyber Mission Force (CMF). CMF has 133 cyber mission teams. These teams work in three areas of cyber warfare. One area protects data and computer systems. Another controls missions that use US cyberpower to harm or deter enemies. The third area designs, builds, and secures US military data and networks.

Cyber warfare has created a new type of soldier. Like all people in the military, cyber warriors go through training. But instead of fighting on a physical battlefield, they use computers to fight in cyberspace. In 2026, the United States was investigating creating a new military branch. It would be called the Cyber Force. It would recruit and train experts in cyber warfare.

Participating in cyber operations such as Cyber Shield 2017 helps US forces develop the skills to counter possible cyberattacks.

CYBER DEFENSE

The best defense against cyberattacks is securing computers. One way to secure data is by encrypting it. Encryption usually relies on difficult math formulas. Updating computer systems makes them more difficult to hack. But more powerful computers and AI make it easier to break encryption code.

CISA keeps track of current cyberattack threats. It then sends alerts to security experts in the government and US companies. This helps security teams find attacks quickly. CISA also provides weekly reports about bugs. It tells security teams how to fix these bugs.

Another weakness in security is people. Security officials add standards for passwords or other security features to computer systems. But some people use passwords that are easy to guess. People are often tricked into clicking on

Selecting a strong, unique password is one line of defense in preventing cyberattacks.

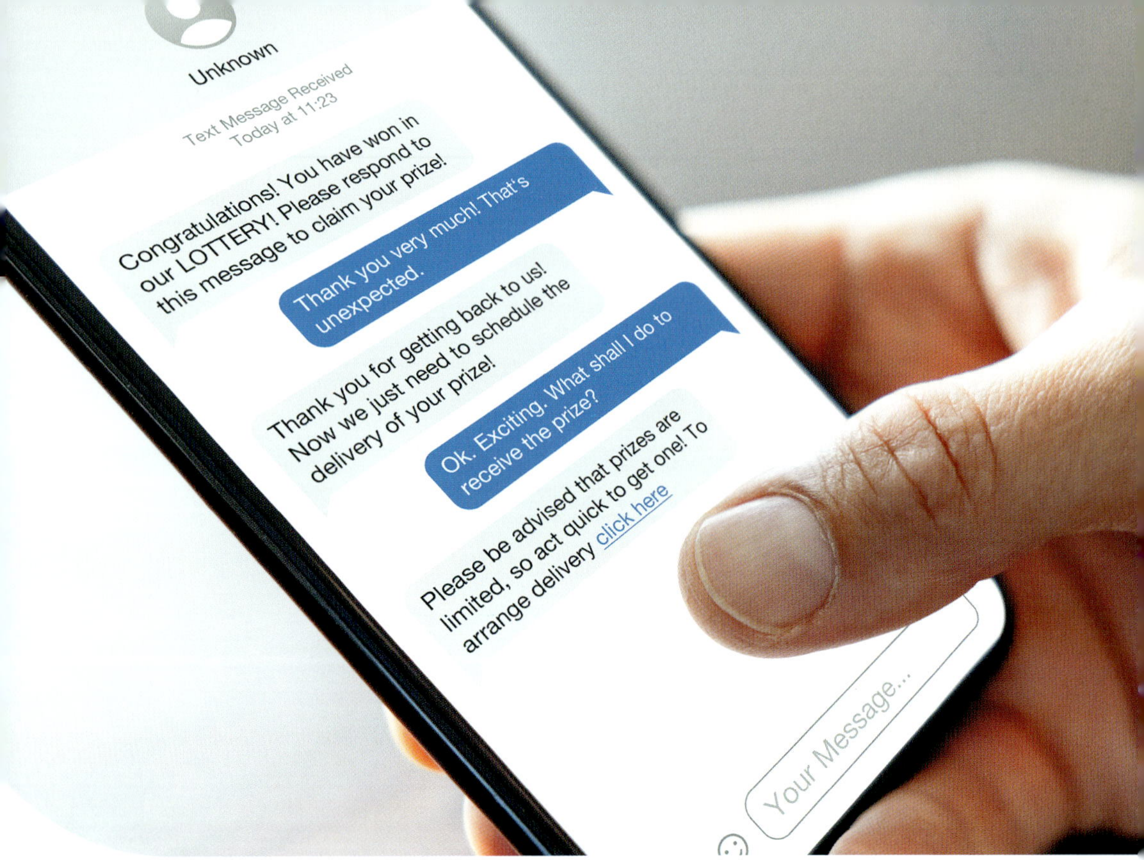

Text messages about package deliveries are common phishing schemes.

phishing links, too. To secure systems, the government and private companies must enforce strict cyber policies.

USERS CAN HELP

Computer users can help find hackers. The Lazarus Group attack was stopped because users were paying attention.

The Lazarus Group tried to send its stolen funds to a bank on Jupiter Street in Manila, Philippines. Jupiter was also the name of an Iranian ship. Someone at the Federal Reserve Bank of New York noticed the name. They discovered five fake transactions. The bank stopped $101 million from leaving the bank. Many more transactions had been planned.

Secrecy adds to the problem. Organizations do not like to admit they were hacked. In 2014, Yahoo had 500 million accounts hacked. Yahoo is a major technology company. It did not announce the hack until 2016. Uber covered up a hack from 2016 to 2022. Uber is a ride-sharing service. It paid a $100,000 ransom to the hackers. The hackers requested payment

Uber, which is based in San Francisco, California, operates in 70 countries and in 15,000 cities.

to delete the stolen data and to keep the attack a secret.

Keeping attacks secret causes problems. Other organizations may not take action to protect themselves. This makes it easier for hackers to break into other systems.

Also, affected customers do not know their data was stolen. The stolen data may be sold to criminals. Kent Walker is head of Google's global affairs. He says, "One of the biggest challenges we encountered was getting companies to go public."[8]

FUTURE OF CYBER WARFARE

The United States is using AI in cyber warfare. AI is used to trace malware. This helps security experts find how it was made and who made it. AI also detects tiny changes in a software program. The change might be a hack.

New technology is changing encryption. Instead of using math, future data may be encrypted with high-level physics. The method uses new discoveries about how tiny particles remain connected over long distances. Information sent this way would be difficult to hack. Canada began testing the technology between Canadian cities in 2025.

Venezuela

In the early hours of January 3, 2026, the United States launched cyberattacks on Caracas, Venezuela. The lights went out in the city. The internet went down. Chinese radar systems failed. Russian air defense systems failed. These cyberattacks helped US forces to achieve their mission. They captured Nicolás Maduro, the leader of Venezuela. They took him to the United States. He was charged with drug trafficking crimes.

Developing new cyber warfare tactics and defenses will continue to be a focus in all branches of the US military for years to come.

Cyber warfare will not stop. Technology is always changing. By building its own cyber warfare program, the military is working to protect the United States from cyberattacks.

GLOSSARY

bots

software programs on the internet that can interact with users

centrifuges

rotating machines that can be used to turn uranium into nuclear fuel

deepfake

a digital body or voice that makes a person appear to be someone else

encrypted

turned into a code to keep information secret

hybrid

something that combines two different methods

negotiate

to attempt to come to an agreement

phishing

sending messages to trick people into revealing information

servers

powerful computers or systems that provide data storage services, such as hosting websites or email

wiretaps

devices used to listen to phone calls

SOURCE NOTES

INTRODUCTION: OPERATION GLOWING SYMPHONY

1. Quoted in Dina Temple-Raston, "How the US Hacked ISIS," *National Public Radio*, September 26, 2019. www.npr.org.

CHAPTER ONE: A NEW WAY TO FIGHT

2. John Arquilla and David Ronfeldt, "Cyberwar Is Coming!" *Rand*, 1993. www.rand.org.

3. Quoted in Joshua Alvarez, "Stuxnet: The World's First Cyber Weapon," *Stanford Center for International Security and Cooperation*, February 3, 2015. https://cisac.fsi.stanford.edu.

CHAPTER TWO: SPYING AND THEFT

4. Quoted in "The Interview: A Guide to the Cyber Attack on Hollywood," *BBC News*, December 29, 2014. www.bbc.com.

5. Quoted in Dina Temple-Raston, "A 'Worst Nightmare' Cyberattack: The Untold Story of the SolarWinds Hack," *National Public Radio*, April 16, 2021. www.npr.org.

CHAPTER THREE: CYBER WARFARE INCREASES

6. Christopher A. Wray, "Director Wray's Remarks at the Munich Security Conference," *Federal Bureau of Investigation*, February 15, 2024. www.fbi.gov.

7. Quoted in Temple-Raston, "How the US Hacked ISIS."

CHAPTER FOUR: WINNING THE CYBER WAR

8. Quoted in Kent Walker, "Transparency in the Shadowy World of Cyberattacks," *The Keyword*, July 19, 2022. https://blog.google/outreach-initiatives/public-policy.

FOR FURTHER RESEARCH

BOOKS

Cynthia Kennedy Henzel, *Advanced Weaponry in the Military.* BrightPoint Press, 2027.

Cynthia Kennedy Henzel, *Space Defense Systems in the Military.* BrightPoint Press, 2027.

Rachel Kehoe, *Cracking the Code: The Human Quest to Uncover Secrets.* Orca Book Publishers, 2025.

INTERNET SOURCES

Katie Lange, "12 Military Innovations That Are Now Everyday Parts of Society," *US Department of War*, November 20, 2025. www.war.gov.

Maggie Miller, "Venezuela Strike Marks a Turning Point for US Cyber Warfare," *Politico*, January 7, 2026. www.politico.com.

Richard Wike et al., "Social Media Seen as Mostly Good for Democracy Across Many Nations, but US Is a Major Outlier," *Pew Research Center*, December 6, 2022. www.pewresearch.org.

WEBSITES

Department of War: Cyber Initiative
www.cybermission.tech

The Department of War Cyber Initiative has information about cyber warfare. It features a game, "Can You Hack It as a Military Cyberspecialist?" to help potential specialists learn more about the field.

National Initiative for Cybersecurity Careers and Studies
https://niccs.cisa.gov/news-events/featured-stories

The National Initiative for Cybersecurity Careers and Studies website provides information about cybersecurity careers with links for students interested in the field.

US Cyber Patriot
www.uscyberpatriot.org

The Air & Space Forces Association sponsors the Cyberpatriot program to inspire K-12 students to learn more about careers in cybersecurity. They offer camps, contests, and opportunities for students to teach others in their community about cybersecurity.

INDEX

IMAGE CREDITS

Cover: Joseph Eddins/DVIDS
5: © DC Studio/Shutterstock Images
7: J.M. Eddins Jr./US Air Force/DVIDS
9: © Mohammad Bash/Shutterstock Images
10: © Digineer Station/Shutterstock Images
13: © DC Studio/Shutterstock Images
15: © Marcio Jose Bastos Silva/Shutterstock Images
17: NASA Images
19: US Department of Energy
21: © Dmitry Kostyukov/AFP/Getty Images
23: © Torsten Pursche/Shutterstock Images
26: © picts/Shutterstock Images
29: © University of College/Shutterstock Images
31: © Media Photos/Shutterstock Images
33: © Michael Vi/Shutterstock Images
34: © ranchorunner/Shutterstock Images
37: © Gino Santa Maria/Shutterstock Images
39: © Ron Adar/Shutterstock Images
40: © bluestork/Shutterstock Images
43: © Tada Images/Shutterstock Images
45: Red Line Editorial
47: Spc. Beverly Roche/US Army/DVIDS
49: Lt. Col. Wayde Minami/US Air National Guard/DVIDS
51: © Garun.Prdt/Shutterstock Image
52: © Kaspars Grinvalds/Shutterstock Images
54: © Alf Ribeiro/Shutterstock Images
57: © Frame Stock Footage/Shutterstock Images

ABOUT THE AUTHOR

Cynthia Kennedy Henzel has degrees in education and geography. She has written more than a hundred books for young people. These include biographies, fiction, and nonfiction on subjects such as geography, science, culture, and history. Henzel worked to bring computers to the classroom in the early 1990s.